可怕的垃圾之路

环保卫士 小猪爱可

小猪爱可阻止垃圾污染

[美] 丽莎·弗兰奇 著　　[美] 巴瑞·戈特 绘　　张玉亮 译

江西科学技术出版社

（本书中文简体版权经由锐拓传媒取得Email:copyright@rightol.com）

版权合同登记号 / 14-2016-0010

图书在版编目（CIP）数据

可怕的垃圾之路：小猪爱可阻止垃圾污染：英汉对照 / (美) 弗兰奇著；(美)戈特绘；张玉亮译.
-- 南昌 :江西科学技术出版社, 2016.8
（小猪爱可讲环保）
ISBN 978-7-5390-5488-9
Ⅰ.①可… Ⅱ.①弗… ②戈… ③张… Ⅲ.①环境保护 – 少儿读物 – 英、汉 Ⅳ.①X-49

中国版本图书馆CIP数据核字(2016)第026821号

国际互联网（Internet）地址：http://www.jxkjcbs.com
选题序号：KX2016083　　图书代码：D16001-101

小猪爱可讲环保
可怕的垃圾之路：小猪爱可阻止垃圾污染

文 / (美) 丽莎·弗兰奇　　图 / (美)巴瑞·戈特　　译 / 张玉亮
责任编辑 / 刘丽婷　　美术编辑 / 刘小萍　曹弟姐
出版发行 / 江西科学技术出版社
社址 / 南昌市蓼洲街2号附1号　　邮编 / 330009
电话 / (0791)86623491　　86639342(传真)
印刷 / 江西华奥印务有限责任公司
经销 / 各地新华书店
成品尺寸 / 235mm × 205mm　　1/16
字数 / 50千　　印张 / 8
版次 / 2016年8月第1版　　2016年8月第1次印刷
书号 / ISBN 978-7-5390-5488-9
定价 / 50.00元（全4册）

赣版权登字-03-2016-7　　版权所有，侵权必究
（赣科版图书凡属印装错误，可向承印厂调换）

It was another grand day
in the town of To-Be.
Eco-Pig napped
in the green apple tree.

未来小镇又迎来了明媚的一天。
小猪爱可懒洋洋地躺在绿色的苹果树
上打着小盹儿。

Yes, all was in order,
so clean and so neat,
from the hairs on each head,
to each home on each street.

从每位居民的头发，到每条
街道上的家家户户，一切都那么干
净、井然有序。

The roses were rosy
and the sky was deep blue.
The songbirds were singing
and the hedgehogs were too!

玫瑰是红艳艳、娇滴滴的，天空是湛蓝湛蓝的。鸟儿在欢快地歌唱，小刺猬也哼着自己的小曲儿。

E.P. was snoring,
dreaming his favorite dreams
of wide-open fields
and clear, bubbling streams.

小猪爱可酣畅地打着呼噜，进入了他最喜爱的梦境。梦中有广阔的原野和翻着水花的小溪。

But something was happening,
something not right!
And when E.P. awoke,
he got quite a fright!

突然，一些意外的事情发生
了！当小猪爱可从梦中醒来，他
大吃一惊！

E.P. could not deny it,
he had to confess.
From river to roadside,
To-Be was suddenly a mess!

　　小猪爱可虽然不愿相信，但也不得不正视眼前发生的一切。从小溪到路边，未来小镇突然间变得一团糟！

There was a trail with flat tires,
a rusty paint can,
and plastic bags by the dozen.
The town was not spick-and-span!

　　道路上七零八落地散落着废弃的轮
胎、生锈的油漆罐和一打打的塑料袋。
未来小镇再也不是一尘不染的了！

9

E.P. said, "I must get to the bottom
of this foul-smelling pile!"
"But how?" cried Louise,
"It goes on for a mile!"

　　小猪爱可发誓："我必须得找到这堆臭烘烘的垃圾的来源！"
　　"怎么找啊？"山羊路易斯大声问道，"垃圾断断续续地散落在道路上，有1600多米长！"

They followed the trash trail
through the town of To-Be.
They went past the park and through
the forest to the edge of the sea.

他们沿着垃圾散落的痕迹穿过未
来小镇，越过公园和森林，一直来到
了海边。

While E.P. got busy
giving garbage the sack,
Louise turned to see
they were still under attack!

　　当小猪爱可正忙着用袋子装垃圾
的时候，山羊路易斯却发现他们依然
在遭受垃圾的侵袭！

"Hey, how's it going?
I'm Pete J. Pollutes.
How about some chewed bubble gum
or some leaky rain boots?"

"嘿，你们在干什么？我是污染大王皮特。要不要来点儿嚼过的泡泡糖或者来几双漏水的雨靴？"

13

"Pardon me, Pete," said E.P.
"But we keep our town neat,
from the leaves on the trees
to the toes on our feet!"

　　"不好意思，打断你一下，皮特，"小猪爱可
生气地说，"我们小镇上每处地方都应该保持清
洁，从树上的叶子到我们的脚尖都应该如此！"

"Why should I care?" Pete replied.
"I'm just passing through."
"Because all of Earth is your home,"
E.P. said. "It's the right thing to do!

　　"我为什么要关心这些呢？"皮特回
答道，"我只不过是路过罢了。""因
为整个地球都是你的家啊，"小猪爱可
说，"这是我们应该做的事情啊！

"Just look at this planet!
What a great work of art!
To protect all this beauty,
we must each do our part!

"看一看我们生活的这个星
球吧！这是一件多么伟大的艺
术品啊！为了保护所有这些美
好的事物，我们必须要做好自
己该做的事情——保护地球！

"We all want clean land.
We all want clean water,
every boy, every girl,
every pig, goat, and otter!"

"我们都希望拥有一方净土，我
们都想要拥有一汪清水。这是每个男
孩、每个女孩、每头小猪、每只山羊
和每只海獭的共同心愿啊！"

17

Olive Otter agreed, "When I snorkel and swim in the peaceful blue sea, your goop, gunk, and garbage choke the life out of me!"

　　海獭奥利夫也随声附和道："当我在平静的蓝色海面之下潜水，畅快地游来游去时，你扔的那些黏糊糊的垃圾却差点儿让我窒息而死！"

"And when I hike in the mountains," added
Louise, "I like to eat what I see.
But I shouldn't eat tin cans,
I'm sure you'll agree."

"当我在山野中徒步远行，"山羊路
易斯也补充说，"我喜欢吃看到的东西。
但是，我可没法吃锡罐，我相信在这一点
上你会认同我的。"

19

"That's a whole lot of hogwash," argued Pete.
"You're just a goat and a pig.
You can't make a difference.
The problem's already too big!"

"怎么这么多废话，"皮特争辩道，"你们不过是山羊和小猪而已。你们什么都改变不了。环境污染本来就已经很严重了！"

"We can all make a difference," E.P. said.
"We can and we do!
Now kindly put down that litter,
and please join my Green crew!

"我们所有人一起努力就能改善环境，"
小猪爱可说，"我们应该身体力行，做力所
能及的事情！现在，拜托你放下那些垃圾，
加入到我的绿色环保团队中来吧！

"Help us clean up this mess,
and let's get on our way.
There are bottles to sort.
It's recycling day!"

"帮我们清理掉这些杂物。我们还有其他事要做，这些瓶子需要进行分类。今天可是回收利用日！"

Pete finally caught on.
"If you all work to keep our home clean,
then I guess my polluting
is selfish and mean!"

污染大王皮特最后终于悔悟了。"如果你
们所有人都在为我们家园的整洁而努力，我觉
得我再去搞污染破坏就显得太自私自利了！"

23

Wiping tears from his eyes,
Pete tossed the boots in the sack.
Then he led that Green crew
as they cleaned their way back.

　　皮特擦去眼角的泪水，将漏水的靴子
放进袋子里。然后他领着绿色环保团队，
沿着他们来时的路清理打扫。

They cleaned at the beach,
down the mountain, in the park,
and all around the school.
They even cleaned an aardvark!

他们清理了沙滩、山脚、公园，也打扫了学校附近。他们甚至还给土豚洗了个澡！

They recycled paper and cardboard,
plastic bottles and bags,
and tin cans and tinfoil.
They threw out a bunch of greasy old rags.

他们将纸张、硬纸板、塑料瓶、袋子、锡罐和锡纸进行了分类回收。为了让未来小镇回归整洁，他们将一大堆抹布用得油腻不堪。

With each sack that he filled,
Pete's smile grew.
He'd made a wonderful discovery—
Earth's beautiful view!

　　看着装满垃圾的一个个袋子，皮特笑得是那么的开心。此时，他有一个惊人的发现——地球上的景色原来是如此的美不胜收！

27

"If you just look around," said E.P.,
"it's easy to care
for the wonder of nature,
for this planet we share."

　　"你只需环顾四周，"小猪爱可说，"很容易就会爱上大自然创造的奇美景象，并且深深地爱上我们共同拥有的这个地球。"

With a smile and a nod,
Pete linked arms with E.P.
as they watched the sun set
over the green apple tree.

　　皮特微笑着点点头，挽着小猪
爱可的胳膊，一起看着落日缓缓地
消失于绿色的苹果树后。

生态学——研究动植物与它们所处环境间关系的一门科学。

绿色环保——与环境或保护环境相关的（事物）。

污染——用人造废弃物破坏环境的行为。

回收利用——将废物、玻璃或易拉罐分类回收，以便再次利用。

你知道吗？

- 公路、森林、公园和沙滩上常常会出现垃圾。垃圾包括：食物包装纸、泡沫塑料杯、纸巾、塑料餐具、瓶瓶罐罐、纸袋和塑料袋、衣物、报纸、杂志、机油桶、油脂、轮胎以及很多其他东西。
- 玻璃瓶分解或消失需要耗费100万年的时间。塑料、泡沫塑料杯以及铝罐需要500年才能分解。
- 美国人每年所扔的塑料瓶足够绕地球4圈。
- 不到0.5千克的报纸可回收制作6个谷类食品包装盒、6个蛋类包装箱或者2000张稿纸。
- 回收利用1个铝罐所节约的能源足够供1个100瓦特的灯泡照明3.5小时。

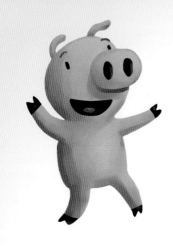

实现环保的更多方式

与你的父母聊聊
你们在家中可以做些什么

1.减少你家产生的垃圾量。

2.将你制造的垃圾随手带走，不要随意丢弃。

3.随手捡起他人丢弃的垃圾。

4.循环利用前期措施：只购买包装可回收利用的物品。

5.回收利用塑料、纸张。

6.回收利用玻璃瓶和铝罐。

7.不要丢弃玩具、书籍或衣物，将这些东西捐赠给医院
或图书馆。